Password Keeper

Keep all your passwords, usernames, and notes in one secure place.

Name

Phone

Location

The best defense is a good offense.

Website	
Email	
Username	
Password	
Question	
Answer	

Security is not a product, but a process

The best defense is a good offense.

Website	
Email	
Username	
Password	
Question	
Answer	

Security is not a product, but a process

The best defense is a good offense.

Website	
Email	
Username	
Password	
Question	
Answer	

Security is not a product, but a process

The best defense is a good offense.

Website	
Email	
Username	
Password	
Question	
Answer	

Security is not a product, but a process

The best defense is a good offense.

Website	
Email	
Username	
Password	
Question	
Answer	

Security is not a product, but a process

The best defense is a good offense.

Website	
Email	
Username	
Password	
Question	
Answer	

Security is not a product, but a process

The best defense is a good offense.

Website	
Email	
Username	
Password	
Question	
Answer	

Security is not a product, but a process

The best defense is a good offense.

Website	
Email	
Username	
Password	
Question	
Answer	

Security is not a product, but a process

The best defense is a good offense.

Website	
Email	
Username	
Password	
Question	
Answer	

Security is not a product, but a process

The best defense is a good offense.

Website	
Email	
Username	
Password	
Question	
Answer	

Security is not a product, but a process

The best defense is a good offense.

Website	
Email	
Username	
Password	
Question	
Answer	

Security is not a product, but a process

The best defense is a good offense.

Website	
Email	
Username	
Password	
Question	
Answer	

Security is not a product, but a process

The best defense is a good offense.

Website	
Email	
Username	
Password	
Question	
Answer	

Security is not a product, but a process

The best defense is a good offense.

Website	
Email	
Username	
Password	
Question	
Answer	

Security is not a product, but a process

The best defense is a good offense.

Website	
Email	
Username	
Password	
Question	
Answer	

Security is not a product, but a process

The best defense is a good offense.

Website	
Email	
Username	
Password	
Question	
Answer	

Security is not a product, but a process

The best defense is a good offense.

Website

Email

Username

Password

Question

Answer

Security is not a product, but a process

The best defense is a good offense.

Website	
Email	
Username	
Password	
Question	
Answer	

Security is not a product, but a process

The best defense is a good offense.

Website	
Email	
Username	
Password	
Question	
Answer	

Security is not a product, but a process

The best defense is a good offense.

Website	
Email	
Username	
Password	
Question	
Answer	

Security is not a product, but a process

The best defense is a good offense.

Website	
Email	
Username	
Password	
Question	
Answer	

Security is not a product, but a process

The best defense is a good offense.

Website	
Email	
Username	
Password	
Question	
Answer	

Security is not a product, but a process

The best defense is a good offense.

Website	
Email	
Username	
Password	
Question	
Answer	

Security is not a product, but a process

The best defense is a good offense.

Website	
Email	
Username	
Password	
Question	
Answer	

Security is not a product, but a process

The best defense is a good offense.

Website	
Email	
Username	
Password	
Question	
Answer	

Security is not a product, but a process

The best defense is a good offense.

Website	
Email	
Username	
Password	
Question	
Answer	

Security is not a product, but a process

The best defense is a good offense.

Website	
Email	
Username	
Password	
Question	
Answer	

Security is not a product, but a process

The best defense is a good offense.

Website	
Email	
Username	
Password	
Question	
Answer	

Security is not a product, but a process

The best defense is a good offense.

Website	
Email	
Username	
Password	
Question	
Answer	

Security is not a product, but a process

The best defense is a good offense.

Website	
Email	
Username	
Password	
Question	
Answer	

Security is not a product, but a process

The best defense is a good offense.

Website	
Email	
Username	
Password	
Question	
Answer	

Security is not a product, but a process

The best defense is a good offense.

Website	
Email	
Username	
Password	
Question	
Answer	

Security is not a product, but a process

The best defense is a good offense.

Website	
Email	
Username	
Password	
Question	
Answer	

Security is not a product, but a process

The best defense is a good offense.

Website	
Email	
Username	
Password	
Question	
Answer	

Security is not a product, but a process

The best defense is a good offense.

Website	
Email	
Username	
Password	
Question	
Answer	

Security is not a product, but a process

The best defense is a good offense.

Website	
Email	
Username	
Password	
Question	
Answer	

Security is not a product, but a process

The best defense is a good offense.

Website	
Email	
Username	
Password	
Question	
Answer	

Security is not a product, but a process

The best defense is a good offense.

Website	
Email	
Username	
Password	
Question	
Answer	

Security is not a product, but a process

The best defense is a good offense.

Website	
Email	
Username	
Password	
Question	
Answer	

Security is not a product, but a process

The best defense is a good offense.

Website	
Email	
Username	
Password	
Question	
Answer	

Security is not a product, but a process

The best defense is a good offense.

Website	
Email	
Username	
Password	
Question	
Answer	

Security is not a product, but a process

The best defense is a good offense.

Website	
Email	
Username	
Password	
Question	
Answer	

Security is not a product, but a process

The best defense is a good offense.

Website	
Email	
Username	
Password	
Question	
Answer	

Security is not a product, but a process

The best defense is a good offense.

Website	
Email	
Username	
Password	
Question	
Answer	

Security is not a product, but a process

The best defense is a good offense.

Website	
Email	
Username	
Password	
Question	
Answer	

Security is not a product, but a process

The best defense is a good offense.

Website	
Email	
Username	
Password	
Question	
Answer	

Security is not a product, but a process

The best defense is a good offense.

Website

Email

Username

Password

Question

Answer

Security is not a product, but a process

The best defense is a good offense.

Website	
Email	
Username	
Password	
Question	
Answer	

Security is not a product, but a process

The best defense is a good offense.

Website	
Email	
Username	
Password	
Question	
Answer	

Security is not a product, but a process

The best defense is a good offense.

Website	
Email	
Username	
Password	
Question	
Answer	

Security is not a product, but a process

The best defense is a good offense.

Website	
Email	
Username	
Password	
Question	
Answer	

Security is not a product, but a process

The best defense is a good offense.

Website	
Email	
Username	
Password	
Question	
Answer	

Security is not a product, but a process

The best defense is a good offense.

Website	
Email	
Username	
Password	
Question	
Answer	

Security is not a product, but a process

The best defense is a good offense.

Website	
Email	
Username	
Password	
Question	
Answer	

Security is not a product, but a process

The best defense is a good offense.

Website	
Email	
Username	
Password	
Question	
Answer	

Security is not a product, but a process

The best defense is a good offense.

Website	
Email	
Username	
Password	
Question	
Answer	

Security is not a product, but a process

The best defense is a good offense.

Website	
Email	
Username	
Password	
Question	
Answer	

Security is not a product, but a process

The best defense is a good offense.

Website	
Email	
Username	
Password	
Question	
Answer	

Security is not a product, but a process

The best defense is a good offense.

Website	
Email	
Username	
Password	
Question	
Answer	

Security is not a product, but a process

The best defense is a good offense.

Website	
Email	
Username	
Password	
Question	
Answer	

Security is not a product, but a process

The best defense is a good offense.

Website	
Email	
Username	
Password	
Question	
Answer	

Security is not a product, but a process

The best defense is a good offense.

Website	
Email	
Username	
Password	
Question	
Answer	

Security is not a product, but a process

The best defense is a good offense.

Website	
Email	
Username	
Password	
Question	
Answer	

Security is not a product, but a process

The best defense is a good offense.

Website	
Email	
Username	
Password	
Question	
Answer	

Security is not a product, but a process

The best defense is a good offense.

Website	
Email	
Username	
Password	
Question	
Answer	

Security is not a product, but a process

The best defense is a good offense.

Website	
Email	
Username	
Password	
Question	
Answer	

Security is not a product, but a process

The best defense is a good offense.

Website	
Email	
Username	
Password	
Question	
Answer	

Security is not a product, but a process

The best defense is a good offense.

Website	
Email	
Username	
Password	
Question	
Answer	

Security is not a product, but a process

The best defense is a good offense.

Website	
Email	
Username	
Password	
Question	
Answer	

Security is not a product, but a process

The best defense is a good offense.

Website

Email

Username

Password

Question

Answer

Security is not a product, but a process

The best defense is a good offense.

Website	
Email	
Username	
Password	
Question	
Answer	

Security is not a product, but a process

The best defense is a good offense.

Website	
Email	
Username	
Password	
Question	
Answer	

Security is not a product, but a process

The best defense is a good offense.

Website	
Email	
Username	
Password	
Question	
Answer	

Security is not a product, but a process

The best defense is a good offense.

Website	
Email	
Username	
Password	
Question	
Answer	

Security is not a product, but a process

The best defense is a good offense.

Website	
Email	
Username	
Password	
Question	
Answer	

Security is not a product, but a process

The best defense is a good offense.

Website	
Email	
Username	
Password	
Question	
Answer	

Security is not a product, but a process

The best defense is a good offense.

Website

Email

Username

Password

Question

Answer

Security is not a product, but a process

The best defense is a good offense.

Website	
Email	
Username	
Password	
Question	
Answer	

Security is not a product, but a process

The best defense is a good offense.

Website	
Email	
Username	
Password	
Question	
Answer	

Security is not a product, but a process

The best defense is a good offense.

Website	
Email	
Username	
Password	
Question	
Answer	

Security is not a product, but a process

The best defense is a good offense.

Website	
Email	
Username	
Password	
Question	
Answer	

Security is not a product, but a process

The best defense is a good offense.

Website	
Email	
Username	
Password	
Question	
Answer	

Security is not a product, but a process

The best defense is a good offense.

Website	
Email	
Username	
Password	
Question	
Answer	

Security is not a product, but a process

Keeping your passwords secure is essential in our digital world. This password keeper is not just a tool; it's your first step toward protecting your online presence. Remember, a strong password is like a fortress, and with this book, you're building the walls. Stay safe, stay organized, and never underestimate the power of good password management!

Made in the USA
Monee, IL
04 November 2026

67386137R10049